xford Science Research Papers 5

The Special Theory of Relativity

A Critical Analysis

L. Essen O.B.E., D.Sc. (NPL retd)

High Hallgate
41 Burofston Park Drive
Gt. Bookham
Surrey

CLARENDON PRESS OXFORD 1971

Oxford University Press, Ely House, London W.1

GLASGOW NEW YORK TORONTO MELBOURNE WELLINGTON

CAPE TOWN SALISBURY IBADAN NAIROBI DAR ES SALAAM LUSAKA

ADDIS ABABA BOMBAY CALCUTTA MADRAS KARACHI LAHORE DACCA

KUALA LUMPUR SINGAPORE HONG KONG TOKYO

Set by E.W.C. Wilkins & Assoc. Ltd.,
and printed in Great Britain by
Lowe and Brydone Ltd, Victoria Road, London

CONTENTS

THE SPECIAL THEORY OF RELATIVITY – A CRITICAL ANALYSIS

1. Introduction

No branch of science has received more public acclaim than the theory of relativity, and few scientists are held in greater esteem than its author, A. Einstein. The theory was accepted by such philosophers of science as B. Russell (1927), de Broglie (1939), and W. Heisenberg (1958). P.W. Bridgeman (1936) accepted the special theory of relativity with some reservations, although he was very critical of the general theory. A. Eddington, who was very influential in popularizing the theory in England, expressed the view in the preface to *Space, time, and gravitation* (1920) that with the theory of relativity Albert Einstein provoked a revolution of thought in physical science; and more recently C. Lanczos (1959) has stated that within the span of a few years the name Einstein attained a lustre that is perhaps unprecedented in the entire history of the human race. In view of such comments it is surprising to find that the theory presents many strange features. Einstein and many other writers have found it necessary to write explanations of the theory, and although the explanations are largely repetitive they sometimes differ in important respects. There are examples of the same author giving, at different times, different explanations of some of the relativity predictions. It is a subject about which writers tend to use more emotive language than is usual in scientific texts. For example, L.B. Loeb and A.S. Adams (1933) state that most of those who attack relativity are either fanatics or so poorly equipped mathematically that they are incapable of understanding or following the processes involved. The reference here to mathematical ability is puzzling because, as we shall see, parts of Einstein's papers that are often criticized involve no mathematics.

Other strange features are the brevity of the introduction in Einstein's paper of 1905, and the omission of any reference to the work of H.A. Lorentz and H. Poincaré, although this was so important that E. Whittaker (1953) in his detailed study attributes the theory entirely to them. Such points of historical interest are also mentioned by G.H. Keswani (1965, 1966) and G.B. Brown (1967). E.G. Cullwick (1959) and H. Dingle (1960, 1967) also criticize the theory. P.M.S. Blackett (1955) quotes a story showing that E. Rutherford attached no importance to it. Wien had been explaining to him that Newton was wrong in the matter of relative velocity and added, 'But no Anglo-Saxon can understand relativity.' 'No,' Rutherford agreed, 'they have too much sense.' This comment was not borne out later, and the theory became accepted wholeheartedly.

Perhaps the strangest feature of all, and the most unfortunate to the development of science, is the use of the thought-experiment. The expression itself is a contradiction in terms, since an experiment is a search for new knowledge that cannot be confirmed, although it might be predicted, by a process of logical thought. A thought-experiment on the other hand cannot provide new knowledge; if it gives a result that is contrary to the theoretical knowledge and assumptions on which it is based then a mistake must have been made. Some of the results of the theory were obtained in this way and differ from the original assumptions. (Essen 1957, 1963a, 1965, 1969.) Einstein himself calls one of the results peculiar, but in fact it must be wrong, since it disagrees with the initial assumptions.

In spite of these unsatisfactory features, and in spite of the fact that attention has been drawn to them, the theory is still generally accepted. It is, of course, taught in universities, usually uncritically, and there are now plans to introduce it into school courses (Rosser 1969). H. Bondi (1967) has written that his ultimate aim is to get special relativity into the primary-school syllabus. In these circumstances it is particularly important that any weaknesses in the theory should be openly discussed and remedied.

A common reaction of experimental physicists to the theory is that although they do not understand it themselves it is so widely accepted that it must be correct. I must confess that until recent years this was my own attitude. I was, however, rather more than usually interested in the subject from a practical point of view, having repeated, with microwaves instead of optical waves (Essen 1955), the celebrated Michelson–Morley experiment, which was the starting point of the theory. Then with the introduction of atomic clocks, and the enormous increase in the accuracy of time measurements that they made possible, the relativity effects became of practical significance. Corrections are already being made to the frequencies corresponding to the atomic transitions employed in the clocks. The correction for the effect known as time dilation is about four parts in 10^{11} for the hydrogen maser and four parts in 10^{13} for the caesium-beam standard. The suggestion has also been made that relativity effects should be considered in the definition of the unit of time. Thus there is an important practical reason for subjecting the prediction of these effects to a critical examination.

Many of the thought-experiments described by Einstein and others involve the comparison of distant clocks. Such comparisons are now made every day at many laboratories throughout the world. The techniques are well known. It seems reasonable, therefore, to consider the thought-experiments in terms of these techniques. When this is done, the errors in the thought-experiments become more obvious. The fact that the errors in the theory arise in the course of the thought-experiments may explain why they were not detected for so long. Theoretical physicists might not have considered them critically from an experimental point of view. But if one has been actually performing such experiments for many years, one is in a more favourable position to detect any departure from the correct procedure. In the existing climate of opinion, one needed to be very confident to speak of definite errors in the theory. Was there not perhaps some subtle interpretation that was being overlooked? A study of the literature did not reveal any, but even so it was the familiarity with the experiments that gave one the necessary confidence to maintain a critical attitude.

The literature sometimes reveals a remarkable vagueness of expression, a lack of a clear statement of the assumptions of the theory, and even a failure to appreciate the basic ideas of physical measurement. Ambiguities are not absent from Einstein's own papers, and various writers, even when advancing different interpretations of the theory, are correct in as much as these interpretations can all be attributed to Einstein. The literature on the subject is extensive, but it will be cited only to show what other arguments have been advanced and how they differ from those given in the present paper. An attempt will be made to make this sufficiently comprehensive to be understood by those who might not have a detailed knowledge of the background, but who care sufficiently about science to feel that the long-standing confusion about the theory should be resolved.

The criticisms made here refer only to the internal contradictions in the theory itself. Einstein and relativitists generally have stressed the need for experimental checks of his theory; but it is equally important that a theory should be self-consistent. If it is shown to be inconsistent then this criticism holds good even if the results of experiment seem to support some of the assumptions or predictions that are made. A short discussion of the experimental evidence is given. There is much more evidence than there was in Einstein's time and it is important that it should be considered critically to check whether the results can be interpreted in the framework of existing theories, or whether they lead to a rational extension of these theories.

2. Foundations of the theory

It was concluded from electromagnetic theory that light is propagated in space with the constant velocity c; and since the earth moves through space the possibility arose of the detection by optical experiments of the effect of this movement on the apparent velocity of light. The failure to detect any first-order effects was explained by Fresnel's drag coefficient, and the Michelson–Morley experiment was carried out in order to determine whether a second-order effect could be observed. The experiment gave a null result, or at least a result much less than that expected. H.A. Lorentz attempted to explain the null result on the basis of electromagnetic theory. Moving charged particles give rise to a magnetic field, thus disturbing the equilibrium of the forces binding the particles together and causing the length of any moving object to be reduced. The requirements of the electromagnetic theory made it necessary for time to change in a similar way, and these assumptions led to the Lorentz transformations. The transformation of coordinates is sometimes a useful device for simplifying the mathematical representation of phenomena. For example, in Newtonian physics an event can be represented by three coordinates x, y, z of position and one t of time. The position of a body moving from the point $x = 0$ at $t = 0$ with velocity v along the x-axis is given by $x = vt$. If we imagine a system of coordinates moving with the velocity v, then in this system the coordinates of the body are x', y', z', t' where

$$x' = x - vt, \tag{1}$$

$$y' = y \tag{2}$$

$$z' = z, \tag{3}$$

$$t' = t, \tag{4}$$

and these are the expressions to be used for a transformation from one set of coordinates to the other. In the Lorentz transformations, one system of coordinates is again moving along the x-axis with the velocity v, and the relationships are such as to make the velocity of light $= c$ in both systems. They are

$$x' = \beta(x - vt), \tag{5}$$

$$y' = y, \tag{6}$$

$$z' = z, \tag{7}$$

$$t' = \beta(t - vx/c^2), \tag{8}$$

where

$$\beta = (1 - v^2/c^2)^{-1/2} \tag{9}$$

This theory was put forward very tentatively and was not generally regarded as being entirely satisfactory. The Lorentz transformations are the basis of the special theory of relativity, but Einstein derived them from two assumptions of a general nature, which he raised to the status of principles.

3. Einstein's principles

Einstein considers the effects of movement between a magnet and a conductor and concludes that the observable results depend only on relative motion. He refers also to the unsuccessful attempts to discover any motion of the earth relative to the 'light medium' and then proceeds to make the two following assumptions (Einstein 1905, p. 38):

(i) The same laws of electrodynamics and optics will be valid for all frames of reference for which the equations of mechanics hold good.

(ii) Light is always propagated in empty space with a definite velocity c, which is independent of the state of motion of the emitting body.

It is not clear whether the second assumption applies to the velocity as a physical process or as a measured quantity depending on length and time; or whether it is relative to the emitting body or to an observer receiving the waves. It is usually interpreted as the velocity relative to an observer. Instead of obtaining values of $c + v$ or $c - v$ for the velocity of light, where v is his own velocity relative to the source, an observer obtains the values c. Thus it appears that there should be no Doppler change of frequency, and yet this effect is known to exist. This interpretation cannot be correct therefore, and in fact, a little later in his paper, Einstein (1905, p. 41) restates the assumption in the following form: 'Any ray of light moves in the "stationary" system of coordinates with the determined velocity c whether the ray be emitted by a stationary or by a moving body.'

The method of determination of the velocity had already been described. A pulse of light is sent from one point to the other and back again and the velocity is found from the time taken for the double journey. The value obtained in this way on classical theory is $c(1 - v^2/c^2)$. The assumption made therefore is that the velocity of light will be c instead of $c(1 - v^2/c^2)$. It is only the second-order term that is assumed not to be present.

4. The basic process of measurement in terms of defined units

Consider now one of the simplest of all measurements, that of a velocity v, expressed as the distance d travelled in a time t. The result is expressed as $v = d/t$. It is possible to define units of any two of the quantities in this expression. In practice, the units of distance and time are defined, and velocity is then measured in terms of these units, that is, in metres per second. If a unit of velocity were to be defined as well, then the value of v could be expressed in two ways, in terms of the unit of velocity and also in terms of the units of length and time. Conflicting results could be obtained. This is all very obvious, but making the velocity of light have the constant value c even to observers in relative motion is comparable to making it a unit of measurement. The definition of the unit of length or of time must be abandoned; or, to meet Einstein's two conditions, it is convenient to abandon both units.

The contraction of length and the dilation of time can now be understood as representing the changes that have to be made to make the results of measurement consistent. There is no question here of a physical theory but simply of a new system of units in which c is constant, and length and time do not have constant units but have units that vary with v^2/c^2. Thus they are no longer independent, and space and time are intermixed by definition and not as a result of some peculiar property of nature.

In several places there is a suggestion that Einstein realized that he was changing the units. For example, he writes (Einstein 1905, p. 49): 'Further we imagine one of the clocks ... to be so adjusted that it marks the time τ.' Since the clocks have initially been made to read 0 at the same instant, the further adjustment constitutes a change in the average rate of the clock. In a popular book (Einstein 1922), he discusses the adjustment in more detail; but mostly he stresses the fact that the clocks used are all identical.

Other writers also appear to realize that a change of units is involved although they may not state this specifically. H. Dingle (1946, p. 23) states that the quantity that is physically important is not l but $l(1 - v^2/c^2)^{1/2}$ and then later (p. 39) that t must be replaced by $t(1 - v^2/c^2)^{1/2}$. He considers moreover that the whole of metrical physics has been constructed on a single unit – that of length. At the time when he wrote, the unit of time was based on astronomy, and it was perhaps reasonable to accept the unit of length as one of the basic units, time being obtained from length and the rotational velocity of the earth, which was the other basic unit. H. Bondi, discussing the subject later (1967, p. 28), more reasonably takes as his units time, and the velocity of light, which he defines as unity. Neither suggest that the theory is to be regarded simply as a change of units. E.G. Cullwick (1959), however, considers that in relativity theory the velocity of light and the dilation of time are simply conventions and this is analagous to regarding them as units of measurement. F.K. Richtmyer and E.H. Kennard (1950) in their excellent volume seem to be in two minds on this question of units. They state first that time scales are set up experimentally in any given frame of reference in such a way as to make the velocity of light the same in opposite directions; but then a few lines later they assume that the same units are used in all frames of reference. It seems to me that these two statements are contradictory.

It might be added here in parentheses that, although for theoretical purposes the units of measurement can be defined in various ways, they are in practice determined by considerations of the permanance, availability, and precision of the standards that are used to define them. The basic units chosen are those of time, length, and mass. In these practical units the velocity of light is a quantity that must be measured, and its value is 299792·5 km/s.

If the theory of relativity is regarded simply as a new system of units it can be made consistent but it serves no useful purpose. Although the relativity effects follow directly from the assumptions, and although there are occasional references to the adjustment of clocks, the usual interpretation is that the effects occur as a result of some physical process even when the units of measurement are not deliberately changed. Einstein (1920, p. 37), for example, states '... As a consequence of its motion the clock goes more slowly than when at rest.' This view is repeated later (Einstein and Infeld 1938) as 'We can well imagine a moving clock having a different rhythm from one at rest.'

5. The synchronization of time and simultaneity

The question of simultaneity occupies a prominent place in Einstein's paper and in most textbooks on relativity, being regarded as an essential feature that explains results that might otherwise appear to be contradictory. Einstein points out that clocks at two distant points can be synchronized only by sending a light signal from one point to the other and back again, measuring the total time occupied, and then assuming that the velocity is the same in the two directions. Now, although this is true, it should be remembered that the very first measurement of the velocity by Römer gave the value in a single direction, and although the accuracy was not high, the result agrees with subsequent values, which correspond to the average velocity in two directions. Within the limits of experimental error it is an established fact that the value is the same in both directions. Einstein then considers the question of simultaneity and shows that events that are simultaneous for one observer are not simultaneous for an observer moving relative to the first. This is, however, a consequence of Einstein's assumption that the measured velocity of light is the same for both of them – that is, of the adoption of the constant value of c as a unit of measurement. There is no such difficulty if this assumption is not made. Nor is there any practical difficulty. Clocks can be synchronized throughout the world with a precision of about $1\mu s$ and their rates can be compared with an error of only one part in 10^{12}. Thus Einstein seems to have over-emphasized these difficulties; and the problem of simultaneity does not seem to play any part in the development of his main equations and results.

6. The Lorentz transformations

Einstein derived the transformations (5)–(9) by calculating the time of a go-and-return journey of a ray of light in a coordinate system moving with a velocity v and making it equal to twice the time of a single journey. The method of derivation is not easy to follow and does not seem to have been adopted by other writers. It must be remembered however that he starts out with postulates that are contradictory in terms of the usual units of measurement, and the problem is to find units that make them conform. It is not a physical theory in the usual sense but simply an assumption that the velocity of light is c in all inertial frames of reference.

Much of the further development of the relativity theory is an application of the transformations. J.L. Synge (1956) expresses the view that the special theory of relativity might be called the theory of the Lorentz transformations, and B. Russell (1927) writes that the whole of the special theory is contained in the transformations. Einstein's theory differs from that of Lorentz only in the method of derivation of the transformations, and that is why it is so important to consider the significance of the postulates and the interpretation of the results. The subsequent mathematical development could be the same in both theories.

7. The dilation of time

One of the most important predictions from the theory is that of the dilation of time, because this is stated in a precise form and related to the actual behaviour of clocks on the earth's surface. The relevant section will be quoted in full (Einstein 1905): 'Further we imagine one of the clocks which are qualified to mark the time t when at rest relatively to the stationary system and the time τ when at rest relatively to the moving system to be located at the origin of the coordinates and so adjusted that it marks the time τ. What is the rate of this clock when viewed from the stationary system?

'Between the quantities x, t, and τ which refer to the position of the clock we have evidently

$$x = vt \tag{10}$$

and
$$\tau = (t - vx/c^2)/(1 - v^2/c^2)^{1/2} \tag{11}$$

Therefore
$$\tau = t(1 - v^2/c^2)^{1/2} \tag{12}$$

$$= t - t\{1 - (1 - v^2/c^2)^{1/2}\} \tag{13}$$

from which it follows that the time marked by the clock (viewed in the stationary system) is slow by $\{1 - (1 - v^2/c^2)^{1/2}\}$ seconds per second or neglecting magnitudes of fourth and higher order – by $\frac{1}{2}(v^2/c^2)$.'

In this paper of Einstein's the introductory statements about the clocks are not precise. One of the clocks is 'qualified' to mark the time t when at rest and τ when it is regarded as moving, and is 'adjusted' to mark the time τ. These statements are consistent with the idea that the rate of the clock is changed, but it is not stated that this is done as a deliberate step. The general interpretation is that the effects occur as a result of some physical effect and not as a deliberate change (see § 4).

Einstein stresses that the clocks are identical and that the results are symmetrical. The symbols t and τ should be interchanged for the two cases, since the clock formerly regarded as stationary is now regarded as moving. This need to interchange the symbols has caused endless argument. H. Dingle (1962, 1967) infers that the Lorentz transformations give the result that one clock goes slower than the other and also that it goes faster than the other. Although this is literally true if the equations as given by Einstein are used, the apparent contradiction is avoided if we interchange the symbols. This seems to be a reasonable thing to do, and Dingle's result has been explained along these lines by a number of writers (Born 1963, Essen 1962, 1963b, 1968, McCrea 1967). Another point that causes confusion is Einstein's expression 'viewed in the stationary system', which subsequently he sometimes omits and sometimes includes. These difficulties are both explained most readily by considering the actual experimental technique for comparing two distant clocks, and by the use of a separate symbol for each measured quantity.

8. Clock comparisons

We measure time by noting the repetitions of some regular process and counting the number on a dial. The word clock denotes usually the combination of a standard vibrator and a counter. We measure the time intervals of a distant clock by receiving timing pulses and counting them on a dial at the receiver. In early experiments, this counting process was actually carried out, but now that it has been established that standard clocks throughout the world keep together to a small fraction of a second per day, it is sufficient to display the received pulse and the pulse from the local clock together on the time scale of a cathode-ray oscilloscope and to note the time difference between them. The principle remains the same however, and there are effectively a standard S and two dials at each station, one dial recording the number of pulses from the local clock and the other the number from the distant clock (see Fig. 1).

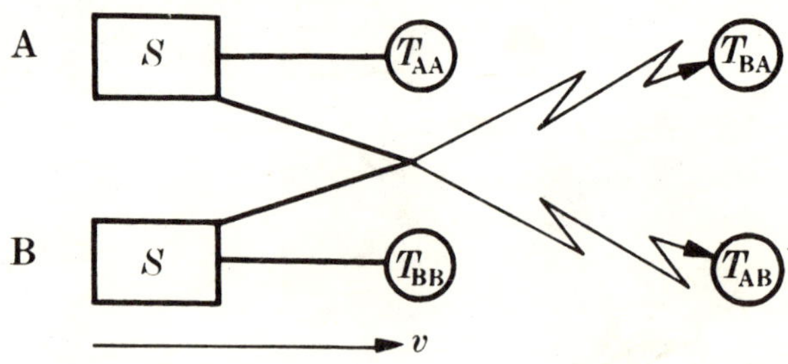

Fig. 1.

The times recorded on the dials can be designated by the suffix AA for the standard at A counted at A, AB for the standard at A counted at B, BB for the standard at B counted at B, and finally BA for the standard at B counted at A. In the previous section, we saw that Einstein was concerned only with the rates of the clocks, and the question of the synchronization of the distant clock did not arise. Time intervals can be represented by Δt, and if the counting of the local signals and those received from the distant clock is started at the same time Einstein's result expressed in these measured quantities is

$$\Delta t_{BA} = \Delta t_{AA}(1 - v^2/c^2)^{1/2}. \tag{14}$$

Einstein's statements that the results are symmetrical and the clocks are identical lead to the following expressions in our terminology:

$$\Delta t_{AB} = \Delta t_{BB}(1 - v^2/c^2)^{1/2} \tag{15}$$

$$\Delta t_{AB} = \Delta t_{BA} \tag{16}$$

$$\Delta t_{AA} = \Delta t_{BB} \tag{17}$$

These equations correspond to Einstein's verbal statements.

Equations (14) and (15) explain immediately Dingle's point: if separate symbols are given to the measured quantities there is no need to interchange them according to which clock is regarded as stationary. There is also no need to add phrases like 'viewed in the stationary system'. The time interval recorded by B's clock is Δt_{BB}, but the time interval given by B's clock when viewed in the stationary system is Δt_{BA}. If the expression 'viewed in the stationary system' is omitted when only two symbols are used for the time interval, Δt_{BA} is confused with Δt_{BB}. If $\Delta t_{BA} = \Delta t_{BB}$ it is clear that eqns (14) and (15) are true only when $v = 0$. Yet many writers seem to regard the qualifying statement as of little importance. Einstein himself sometimes omits it. Some writers, for example C. Møller (1952), omit it even at the beginning of the discussion, but many writers retain it, as Einstein does, until the thought-experiment described in the next section. The problem is linked with the old question of whether the relativity effects are real or apparent. Both Δt_{BB} and Δt_{BA} are real in that they are both measured quantities, but the change in the rate of the clock according to the above equations is only apparent.

It is very important to notice that the equations can be true only if $v = 0$ or if in a time $\Delta t_{AA} = \Delta t_{BB}$ more pulses are transmitted from a clock than are received at the other station. In a manner that is not explained some of the pulses are lost. This is not an impossible situation since there is no complete understanding of the nature of light or of how it is propagated. However, a much more reasonable assumption, and one that is often made, is that all the pulses that are transmitted arrive in the time Δt_{AA} (apart from the Doppler effect, discussed in § 9). There is only one way of making this assumption conform with the equations. This is to assume that each observer has two clocks, one for measuring time intervals between events at rest with him, and one for measuring events in motion relative to him. The second clock must be deliberately adjusted according to the velocity of the moving events. Everything is then consistent but the adjustment constitutes merely a change of units (see § 4).

9. The Doppler effect

The equations (14) and (15) in § 8 appear to be incorrect because, apart from any other considerations, they do not contain the first-order Doppler effect. This is explained, however, by Einstein's definition of time (Einstein 1905, p. 40): 'The time of an event is that which is given simultaneously with the event by a stationary clock located at the place of the event.' In order to avoid the need to allow for the travel time of signals, he, as it were, fills the space with synchronized clocks and reads whichever is at the position of the event. In practice, the travel time must be taken into account. If there is no relative motion between the transmitter and receiver the travel time is constant and does not affect measurements of the rates of clocks; but if the transmitter and receiver are in relative motion the varying travel time gives rise to the Doppler change of frequency – or rate of reception of the time signals. Einstein derives a Doppler formula, but it can be obtained quite simply from eqn (14), if we remember that Δt_{BA} represents the number of pulses received. The same relationship will therefore hold for the frequency of a received signal,

$$f_{BA} = f_{AA}(1 - v^2/c^2)^{1/2}, \tag{18}$$

and from eqn (17)

$$f_{BA} = f_{BB}(1 - v^2/c^2)^{1/2}. \tag{19}$$

The ordinary Doppler formula for a fixed observer and moving source is

$$f' = f/(1 - v/c), \tag{20}$$

so that f_{BB} should be replaced by $f_{BB}/(1 - v/c)$. We thus have

$$\begin{aligned} f_{BA} &= f_{BB}(1 - v^2/c^2)^{1/2}/(1 - v/c) \\ &= f_{BB}(1 + v/c)^{1/2}/(1 - v/c)^{1/2}, \end{aligned} \tag{21}$$

which agrees with Einstein's result.

10. The clock paradox

The argument about the clock paradox has continued interminably, although the way the paradox arose and its explanation follow quite clearly from a careful reading of Einstein's paper. Einstein's result, given in § 7, is 'that the time marked by the moving clock viewed in the stationary system is slow'; and it follows from the assumption of symmetry that:

(i) clock B viewed in A is slower than clock A.

(ii) clock A viewed in B is slower than clock B.

There is no contradiction here, but in the next section of his paper Einstein omits the expression 'viewed in A' and reaches the conclusion that clock B is slower than clock A, which contradicts the postulate of symmetry. Einstein calls the result peculiar, but it does not follow from the thought-experiment if this is correctly performed. According to Einstein, one clock A moves along a line AB with velocity v to point B. When it reaches B its reading is slow compared with that of the clock which has remained at B and was initially synchronized with A. The result is said to hold good for any polygonal line and also when the points A and B coincide. Finally it is assumed to be valid for a closed curve, so that if a clock makes a round trip from A it will be slower on its return than a clock that has remained at

A. Although the clock must be accelerated during this journey, no allowance is made for any effect of the acceleration – which indeed is not mentioned.

Consider how the experiment should be performed. At A there are two dials, one recording the pulses from a clock at A, and one recording those from the travelling clock, which will be denoted as B. Then, if the effect of accelerations is neglected, the time intervals recorded at the end of the journey will be those given by eqn (14), the counting on both dials being started at the moment of departure and stopped at the moment of return. The experiment does not give any value for the reading on B's clock. However, to complete the experiment an observer with B will have two clock dials, one recording the pulses from his clock B and the other those from clock A, and he will obtain the result given by eqn (15). Einstein obtains an incorrect result simply by omitting the expression 'viewed in the stationary system'. He makes this omission right at the beginning of his thought-experiment, when the clock moves linearly from A to B. The result has therefore nothing to do with accelerations, both because it was obtained first for linear motion only, and because in the round trip no allowance is made for the effect of acceleration.

The experiment is often expressed in the dramatized form of two twins, one of whom returns from a round trip younger than his brother; and in this form it has received wide publicity. Even in *The Times* (25 November 1960) there appeared the headline: 'Prolonging life by space travel'. This description of time dilation is consistent with the idea that it is simply a change of unit. Such a change would naturally affect the rate of all processes. A traveller could age half as rapidly in terms of a clock that was adjusted to go at half the normal rate, but this would not affect his physical condition.

By omitting the expression 'viewed in the stationary system', Einstein is making a new, additional assumption of some kind. The only way symmetry can be preserved is by the assumption that we are concerned simply with a change of units. This explanation is rejected in many of Einstein's statements, and in this case the dilation must be an actual change in the clock's rate as a result of its motion. In this case, symmetry no longer exists and the relativity postulate is not valid. If the effect is an actual change in the rate of an atomic clock – which is quite a rational idea – it cannot be extended to the time duration of other processes, such as the ageing of an individual.

11. Explanations in terms of gravitational field

This 'paradox' result must have been criticized many years ago, because in 1918 Einstein published a strange paper consisting of a dialogue between a critic and a relativitist. The critic repeats the treatment given in Einstein's paper but adds that, according to the principle of relativity, the results must be symmetrical, so that it can be concluded not only that clock B goes slower than clock A but also that clock A goes slower than clock B. The relativitist agrees that this must be wrong and explains that the result rests on the fact that the two observers are not symmetrical and view the events in different ways. According to A, B is accelerated and then travels with velocity v to a distant point. It is then acted on by an accelerating force until its velocity is v in the opposite direction. It moves back to A and is decelerated to rest. Clock B is slow on its return. In order to perform the journey, it is acted on by accelerations, which do not, however, affect its rate. From B's point of view, a gravitational force acts long enough to impart a velocity v to A. At the distant point, another gravitational force reverses the velocity of A and causes it to travel back to B. Clock A loses relative to B because of its travel with the velocity v, but during its reversal of direction it is at a different gravitational potential from B, and this causes it to gain twice as much time as is lost due to the velocity, so that both observers agree that clock B is slower than clock A. Einstein describes this thought-experiment as the complete explanation of the paradox; however, in my view it is not only illogical but contains two specific errors made in the course of the 'experiment'. It is illogical to suggest that a result obtained on the basis of the special theory is correct but is a consequence of a completely different theory developed some years later. It is also illogical to assume that accelerations have no effect – as he does in A's picture of the events – and then to assume that gravitation, which in the general theory is assumed to be equivalent to acceleration, does have an effect.

The experiment described as A's picture is exactly the same as that described in the special theory and reproduced in § 10. It contains the same 'experimental error' of omitting the phrase 'viewed in the stationary system'. Although the experiment is viewed in two different ways there is only one cycle of events. If the hypothetical gravitational field causes an actual change in the rate of A's clock then this change must obviously be noticed by A – but it is not noticed according to Einstein's description.

Einstein's argument has not been widely adopted, although it is given by R.C. Tolman (1934), who first describes time dilation as a symmetrical effect, and then gives the unsymmetrical result for the round-trip experiment. He explains it in terms of gravitational potential, giving the actual calculations. W.G.V. Rosser (1964) also gives the tw explanations. He does not claim to have resolved the paradox but merely to have obtained the same answer in two different ways.

The explanation given by M. Born (1924) in his well-known book implies that the result of the round-trip experiment is a consequence of the acceleration, although the nature of the effect is not explained, and it has been assumed earlier that accelerations have no effect.

12. Explanations in terms of the special theory of relativity

There is a fairly widely held opinion that the asymmetry of the two clocks must be responsible for an unsymmetrical result, so that there is no paradox. The result, however, is obtained from the special theory alone without introducing any effect due to acceleration or gravitation. The assumption is made that all the timing pulses emitted by the moving clock are received by the stationary clock during the journey. This assumption appears to be so reasonable and obvious that it is sometimes made implicitly. But it seems to me that the only way to obtain a symmetrical result without changing the rates of the clocks is to assume the non-arrival of some of the pulses in the time concerned (cf. § 8). Thus the assumption that they all arrive is contrary in my view to Einstein's time dilation prediction as outlined in § 7. It is equivalent to equating Δt_{BA} with Δt_{BB}, or to dropping expressions such as 'viewed in the stationary system'.

H. Bondi (1957) gives a particularly clear version of the thought-experiment, which at the same time provides an interesting way of obtaining the Doppler formula.

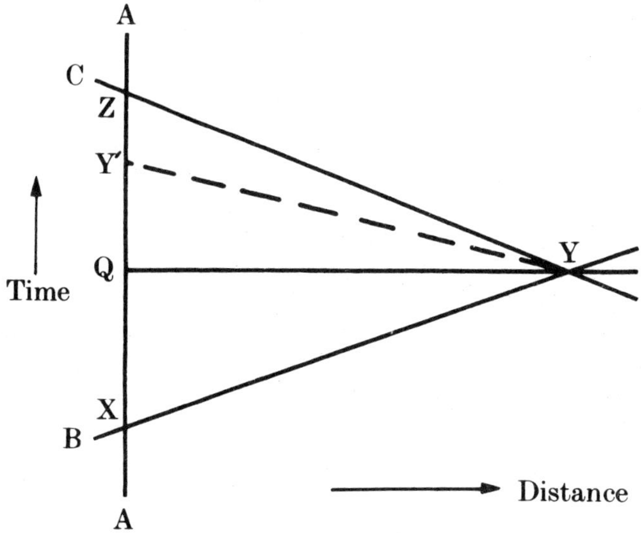

Fig. 2.

In Bondi's nomenclature, A (in Fig. 2) is a stationary observer and B passes him with a velocity v at time X. At time Y, B is passed by C moving with velocity-v. C reaches A at time Z. This construction avoids the introduction of accelerations, and enables an explanation of the paradox to be given in terms of the special theory. The observers all have identical clocks sending out pulses at time intervals h apart. Q is the mid-point between X and Z. It is assumed that A's clock reads t_0 at Q and $2t_0$ at Z; and that B's clock reads t_1 at Y when C's clock is synchronized with it. C's clock reads $2t_1$ at Z. Bondi introduces a factor k such that pulses sent out by B at intervals of h arrive at A at intervals kh measured on A's clock. The pulses from C arrive at intervals of h/k. The distance from A of the point at which B and C cross is vt_0 measured in A's time, and the time taken by a pulse of light to travel from this point to A is vt_0/c. Hence $QY' = vt_0/c$, and A receives pulses from B for the time $t_0 + vt_0/c$ and pulses from C for the time $t_0 - vt_0/c$. The number of pulses emitted by both in time t_1 is t_1/h. Bondi assumes that all the pulses emitted by B arrive at A in the time $t_0 + vt_0/c$ and all those from C in the time $t_0 - vt_0/c$ so that

$$t_0 (1 + v/c) = kh \times t_1/h = kt_1 \tag{22}$$

and

$$t_0 (1 - v/c) = h/k \times t_1/h = t_1/k. \tag{23}$$

By multiplication of eqns (22) and (23), we get

$$t_0^2 (1 - v^2/c^2) = t_1^2, \tag{24}$$

and by division of eqn (22) by eqn (23),

$$(1 + v/c)/(1 - v/c) = k^2. \tag{25}$$

Now the clocks are stated to be identical, and since they are recording over the same interval of time between X and Z, I can interpret this only as meaning that $t_0 = t_1$, which contradicts eqn (24). Let us therefore retain the assumption that $t_0 = t_1$ and assume in accordance with the discussion in § 8 that, of the pulses emitted in time t_1, only the fraction αt_1 are observed by A in the time $t_0 (1 + v/c)$. Equations (22) and (23) then become

$$t_0 (1 + v/c) = \alpha k t_1 \tag{26}$$

and

$$t_0 (1 - v/c) = \alpha t_1/k, \tag{27}$$

and multiplying (26) and (27) and putting $t_0 = t_1$, we get

$$\alpha = (1 - v^2/c^2)^{1/2}. \tag{28}$$

Equation (25) remains true, and thus the experiment constitutes an elegant way of finding the relativity formulae for time dilation and the Doppler effect. If four clock dials had been used, as in our description of the time comparison, we would have found $t_1 = \Delta t_{BB}$, whereas $\alpha t_1 = \Delta t_{BA}$, and it is only this latter quantity which can be measured by A. If measurements of A's time (in Fig. 2) were made by B and C the symmetrical result

$$\alpha^2 t_0^2 = t_1^2(1 - v^2/c^2) \tag{29}$$

would be obtained. Bondi explains in general terms that the unsymmetrical result is a consequence of the accelerations that must be present in an actual experiment.

A full treatment of the clock paradox is given by G.J. Whitrow (1961). He places particular stress on the fact that the time dilation is an apparent effect and is symmetrical, stating that just as B's clock seems to A to run slow, so A's clock seems to B to run slow. In the round-trip thought-experiment he omits the word 'seems' and states that B's clock will lag behind A's when they meet again. It is pointed out that there is an absolute difference between A and B in their respective relations with the universe as a whole, but it is not indicated how this changes the rates of the clocks.

Many writers are satisfied that the result follows directly from the special theory, and they obtain it by assuming that all the pulses arrive in the time occupied by the journey. C.G. Darwin (1957) states: 'There is no doubt whatever that the accepted theory of relativity is a complete and self-consistent theory, and it quite definitely implies that a space traveller will return from his journey younger than his stay-at-home twin brother.' Like many others, he employs the Doppler formula (21) instead of eqns (14), (15). This is more in keeping with the actual experiment than is Einstein's own treatment, but it is not strictly necessary, since the frequency shifts due to the ordinary Doppler effect cancel out in the course of the experiment.

H. Dingle's treatment of the problem deserves special mention because he was the first to point out, at least in the modern phase of the controversy, that the clock-paradox result was an actual mistake in Einstein's paper (Dingle 1956). He attributes the mistake to the fact that the Lorentz transformations in two different directions do not commute, whereas in my view the mistake is made in the first leg of the journey, before the question of commutation arises. In a later paper (1957), he gives what he calls a complete solution of the problem. He makes the same assumption that every flash emitted by one observer is received by the other as do those who obtain the unsymmetrical result, but in addition he assumes that the result is symmetrical. His analysis, however, leads not only to the result he gives but also to the result $c - v = c + v$ and thus can be valid only when $v = 0$. In his later papers (Dingle 1960, 1962, 1963, 1967), he argues more generally that if Einstein's arguments are valid the result must be symmetrical, and he uses the Lorentz transformations to obtain the result that the moving clock is both faster and slower than the stationary one. Nevertheless he made an important contribution by keeping the controversy open and making some sound general criticisms of the theory.

13. Consequences of the paradox

It has been shown here that the paradox does not follow from the theory and is the result of confusing the quantities being measured in a thought-experiment. The result is, however, generally accepted, and it is therefore important to consider the implications of this acceptance. There are two possibilities:

(i) The clock rates are changed when observations are made on a moving body.

(ii) It is assumed that there is an actual time dilation, which has a magnitude $t(1 - v^2/c^2)^{1/2}$ arising in some way that is not explained. There is no longer any symmetry between the results of two observers.

It would follow from the first possibility that the Lorentz transformations have no physical importance, but this is contrary to experience. The second possibility is therefore the more reasonable. Thus in accepting this result Einstein abandoned his relativity postulate and returned to the position stated earlier by Lorentz.

14. Rotational effects

It is important to consider rotational effects because in practice it is easier to conduct laboratory experiments on bodies in rapid rotation than on those having a high linear velocity. In his introduction to the general theory, Einstein (1916) describes a thought-experiment in which there is one clock at the circumference of a rotating disc and one at the centre. He writes, 'By a familiar result of the special theory of relativity the clock at the circumference – judged from K – goes more slowly than the other because the former is in motion and the latter is at rest. An observer at the common origin of co-ordinates capable of observing the clock at the circumference by means of light would therefore see it lagging behind the clock beside him. As he will not make up his mind to let the velocity of light along the path in question depend explicitly on the time, he will interpret his observations as showing that the clock at the circumference 'really' goes more slowly than the clock at the origin.'

In this paragraph Einstein starts with the correct experimental technique, and his observer has two readings, one of his own clock, and one of the distant clock obtained by receiving light signals from it.

To complete the experimental arrangement, an observer at the circumference will also have two readings, and according to Einstein's result (§ 7) the readings will be symmetrical unless some factor is introduced to allow for any effect due to the rotation, which is not done. The symmetrical result, as stated in § 8, can be obtained only by a deliberate change of units or a loss of some of the pulses. Einstein is not able to accept this consequence however, and states that the clock at the circumference really goes more slowly. Thus he makes the further implicit assumptions that all the pulses arrive, and that there is some physical change in the clock that is rotating. As in the clock-paradox thought-experiment, it is implied that the result follows from the time-dilation prediction, but in fact an additional assumption is made which contradicts the relativity principle.

15. Gravitational effect

It is not proposed to deal here in any detail with the general theory of relativity, but it is important to examine one result that is of practical importance in the application of atomic clocks. This is the frequency shift due to gravitational potential.

Einstein (1911) considers two frames of reference, one being accelerated uniformly at γ cm/s^2 relative to the other. Light of frequency f_0 is emitted from a source in the frame not being accelerated, at a time when there is no relative velocity between the two frames. If the light has travelled a distance h at velocity c when it reaches the accelerated system, the latter will then have a velocity $\gamma h/c$; and therefore in accordance with Doppler's principle the observed frequency will be

$$f = f_0 (1 + \gamma h/c^2).$$

Now if acceleration is assumed to be equivalent to a gravitational field, the same result holds if the system is not accelerated but provided with a uniform gravitational field. The gravitational potential difference ϕ can be substituted for γh, which gives

$$f = f_0 (1 + \phi/c^2).$$

Einstein then argues that these equations seem to assert an absurdity, since they imply that all the transmitted pulses of light cannot be received; but, he says, the answer is simple. 'Nothing compels us to assume that the clocks in different gravitational potentials must be regarded as going at the same rate.' The rate of the clock must change according to the gravitational potential.

In the argument above there is no suggestion of a physical theory; it consists simply of a number of assumptions. The Doppler effect is a well-understood phenomenon observed when the source and receiver of waves are in relative motion. It is a consequence of the motion and is quantitatively in linear proportion to the velocity. There is no logical reason, nor any physical explanation, why it should be in any way caused by the acceleration that produces the motion. There is, therefore, no logical reason why it should be caused by the gravitational potential, which is assumed to be equivalent to the acceleration. In any case, the assumptions up to this point lead to an apparent effect. Einstein then argues that this is absurd and that it must be the clocks that change their rate in a gravitational field. Again, there is no logical reasoning in this. He simply does not accept the first result and makes another assumption to give a result that seems to be more plausible.

16. A critical examination of the foundations of the theory

In the previous sections, it has been suggested that the theory consists of a number of contradictory assumptions and adds nothing significant to that of Lorentz. It might be useful therefore to re-examine the foundations that were described briefly in §2 in the manner in which they are usually presented.

An essential step in the development of scientific theories is the generalization of experimental results, but it is important to distinguish between the facts established by experiment and the extrapolations from these facts. There is a tendency to ignore the complex and inescapable environment in which experiments are carried out on the surface of the earth, where there is a gravitational field and a centrifugal acceleration (Essen 1963c). All the experiments leading to the electromagnetic theory were carried out in this environment. It is known that any effects of the field and acceleration must be very small, and it is therefore reasonable to assume that the constant c in the theory represents the velocity of waves in 'free space', although it is strictly the velocity in the earth environment. In the same way, the Michelson–Morley experiment was carried out on the earth's surface and could not give any direct information about the velocity of light in free space. The result showed that the velocity in different directions on the earth's surface was the same, a result that is in agreement with a strict interpretation of electromagnetic theory. There was no discrepancy between the experiment and the theory, but only between the experiment and the extrapolation of the theory to free space. A legitimate question to ask would have been: 'Why does the earth environment differ from free space in respect of the propagation of light?'

Lorentz based his explanation of the result of the Michelson–Morley experiment on the assumption that a moving charge gives rise to a magnetic field. The statement is often made that this effect was proved experimentally by H.A. Rowland, but this also is an extrapolation from the experimental results. Rowland and the many workers who repeated the experiment, often with conflicting results (Sivadjian 1953), all used charged discs in rapid rotation. The fields produced were on the borderline of what could be measured, but the experiments probably established that a charged plate rotating relative to an earthed plate gives rise to a magnetic field. It is interesting to recall that there is a similar failure to distinguish between uniform and rotational movement in the relativity theory. Einstein assumes that the dilation of time found for uniform relative velocity applies also to rotation.

If it is assumed from Rowland's result that a uniformly moving charge is accompanied by a magnetic field, the question arises, 'Moving relative to what?', and to answer this question we cannot do better than consider M. Faraday's original discussion of the problem (Whittaker 1949). He first considered two charged spheres moving relative to each other; and then he extended the discussion to a single sphere moving in a room, because in his picture of the process lines of force from the sphere must terminate somewhere in the room. Thus the motion was relative to the walls of the room, or in electrical terms 'earth'. The Lorentz effects of length contraction and time dilation are therefore physical effects arising from the movement of a body relative to an electrical 'earth'.

In considering the earth environment of gravitational field and acceleration, I am not suggesting definitely that they are responsible for the Lorentz effects, but only that their possible influence should be considered. It is possible that electromagnetic theory is incomplete in other, more important ways. G.B. Brown (1958), for example, has pointed out that, if it is assumed that the forces between electric charges are established with a velocity c and not instantaneously, the force between moving charges contains a term v^2/c^2. E.G. Cullwick (1959) has suggested that there is now experimental evidence that Maxwell's theory is incomplete.

17. Relevant experiments

It is a common view that the special theory of relativity is well supported by experimental evidence, although this may not be true of the general theory. For example, W. Heisenberg (1958) stresses the experimental support and concludes that in consequence the theory belongs to the firm foundations of modern physics and cannot be disputed. It may be surprising, therefore, to find that a more critical examination of the experiments and the experimental conditions suggests that there is no experimental support for the theory. The theory differs from that of Lorentz mainly in that it is a symmetrical effect. It is one of its basic postulates that two observers in relative motion will obtain the same results from physical measurements, but, as Cullwick (1959) has pointed out, no experiment of this kind has ever been performed. The theory applies also to inertial systems, that is, to an environment in which there are no acceleration forces; again, no experiment has ever been carried out in such an environment (Essen 1963a).

The experiments of the Michelson–Morley type cannot be taken as supporting the theory, because the theory was developed in order to explain the null result that was obtained. This does not diminish the importance of the result. Another result often quoted in support of the theory is the variation of the life-time of mesons, the life-time being greater the greater the velocity of the mesons. Again it is an important result, but it cannot be regarded as a confirmation of relativity theory, although it could perhaps be taken to support the theory of Lorentz. The same applies to the experiment of H.E. Ives and G.R. Stillwell (1938), which is often quoted in support of the theory of relativity. They accelerated hydrogen ions and examined the spectra at different velocities. As with all such experiments the great problem was to eliminate the first-order Doppler effect. The displacement of the lines due to this was of the order of 2mm, whereas that due to the second-order effect was only 0·005mm. The reading accuracy was 0·001mm. The spectra were observed with the spectrometer in line with the beam both directly and after reflection. In addition to the first-order shift, a shift of the centre of gravity of the spectrum was observed when the experimental conditions were carefully chosen to avoid interference from lines of the molecular spectrum. The authors concluded that the change of frequency of a moving light source predicted by the Larmor–Lorentz theory was verified. H.E. Ives (1951) was very critical of the theory of relativity and regarded the principle of the constancy of the velocity of light as a paradox.

The increase of mass with velocity was predicted for the case of charged particles directly from electromagnetic theory before the advent of relativity theory and was confirmed experimentally by Kaufmann. The effect is particularly important for the theory of atomic orbits and the design of atomic accelerators but it cannot be claimed justly that the effect is a prediction of relativity theory. The experiments of D.C. Champeney and P.B. Moon (1961) and of D.C. Champeney, G.R. Isaak, and A.M. Khan (1965) are of especial interest. They utilize the extremely sharp resonance obtained from the Mössbauer effect. When the source is at the centre of a rotating disc and the absorber is on the periphery, a frequency shift is observed; but when the source and absorber are at opposite ends of a diameter there is no shift. The results are stated to be in conformity with relativity theory, but the simplest interpretation is that the shift is due to a difference between the accelerations of the source and absorber, and not their relative velocity.

An experiment to determine the effect of gravity on gamma radiation was performed by R.V. Pound and G.A.J. Rebka (1960) and repeated by R.V. Pound and J.L. Snider (1965). The object was to check Einstein's postulate that no local experiment could distinguish between the effects of a gravitational field and the effects of a uniform acceleration of the laboratory with respect to an inertial frame. The source and absorber were separated by a height of 23m. The equivalent acceleration acting through this distance would give a velocity change $\Delta v = gh/c$, and it was therefore predicted that if the source were given an upward velocity of gh/c the effect of g would be cancelled. The result confirmed the prediction, but the authors point out that no strictly relativistic concepts are involved. The experiment is unable to distinguish between frequency changes and velocity changes. The authors conclude that an experimental comparison of clocks at different gravitational potentials would be useful.

18. Conclusions

A critical examination of Einstein's papers reveals that in the course of thought-experiments he makes implicit assumptions that are additional and contrary to his two initial principles. The initial postulates of relativity and the constancy of the velocity of light lead directly to length contraction and time dilation simply as new units of measurements, and in several places Einstein gives support to this view by making his observers adjust their clocks. More usually, and this constitutes the second set of assumptions, he regards the changes as being observed effects, even when the units are not deliberately changed. This implies that there is some physical effect even if it is not understood or described. The results are symmetrical to observers in relative motion; and as such can only be an effect in the process of the transmission of the signals. The third assumption is that the clocks and lengths actually change. In this case the relativity postulate can no longer hold.

The first approach, in which the units of measurement are changed, is not a physical theory, and the question of experimental evidence does not arise. There is no evidence for the second approach because no symmetrical experiment has ever been made. There is no direct experimental evidence of the third statement of the theory because no experiments have been made in an inertial system. There are experimental results that support the idea of an observed time dilation, but accelerations are always involved, and there is some indication that they are responsible for the observed effects.

19. Acknowledgements

The author's views expressed in this paper were reached slowly and after a long consideration of the initial and supporting papers. The problems were discussed with colleagues and by correspondence with a number of theoretical physicists who were interested. All this discussion and correspondence contributed to what is believed to be an explanation of the difficulties associated with the theory; and its value is gratefully acknowledged even though the opinions of the correspondents were often strongly opposed to those expressed here.

REFERENCES

BLACKETT, P.M.S. (1955). *Yb. phys. Soc.*

BONDI, H. (1957). *Discovery* **18**, 505.

(1967). *Assumption and myth in physical theory*, p. 28. Cambridge University Press.

BORN, M. (1924). *Einstein's theory of relativity*. Methuen, London.

(1963). *Nature, Lond.* **197**, 1287.

BRIDGEMAN, P.W. (1936). *The nature of physical theory*. Princeton University Press.

BROGLIE, L. DE (1939). *Matter and light*. Allen and Unwin, London.

BROWN, G.B. (1958). *Sci. Prog, Lond.* **46**, 15.

(1967). *Bull. Inst. Phys., Lond.* p. 71.

CHAMPENEY, D.C. and MOON, P.B. (1961). *Proc. phys. Soc.* **77**, 350.

CHAMPENEY, D.C., ISAAK, G.R. and KHAN, A.M. (1965). *Proc. phys. Soc.* **85**, 583.

CULLWICK, E.G. (1959). *Bull Inst. Phys., Lond.* **10**, 52.

DARWIN, C.G. (1957). *Nature, Lond.* **180**, 976.

DINGLE, H., (1946). *The special theory of relativity*, p. 23, Methuen, London.

(1956). *Nature, Lond.* **177**, 782.

(1957). *Nature, Lond.* **180**, 1275.

(1960). *Philosophy Sci.* **27**, 233.

(1962). *Nature, Lond.* **195**, 985.

(1963). *Nature, Lond.* **197**, 1248.

(1967). *Nature, Lond.* **216**, 119.

(1968). *Nature, Lond.* **217**, 19.

EDDINGTON, A. (1920). *Space, time, and gravitation*. Cambridge University Press.

EINSTEIN, A. (1905). Translation in *The principle of relativity*, pp. 38, 41, 49, Methuen, London (1923).

(1907). ibid.

(1911). ibid.

(1916). ibid.

(1918). *Naturwissenschaften* **48**, 697.

(1920). *Relativity*, p. 37. Methuen, London.

(1922). *The meaning of relativity.* Methuen, London.
EINSTEIN, A. and INFELD, L. (1938). *The evolution of physics.* Cambridge University Press.
ESSEN, L. (1955). *Nature, Lond.* **175**, 793.
(1957). *Nature, Lond.* **180**, 1061.
(1962). *Proc. R. Soc.* **A270**, 312.
(1963a). *J. Instn elect. Engrs* **9**, 389.
(1963b). *Air, space, instruments* (ed. S. Lees), p. 194. McGraw Hill, New York.
(1963c). *Nature, Lond.* **199**, 684.
(1965). *Proc. phys. Soc.* **86**, pt. 3, 671.
(1968). *Nature, Lond.* **217**, 19.
(1969). *Proc. Instn elect. Engrs.* **115**, 1853.

HEISENBERG, W. (1958). *Physics and philosophy.* Allen and Unwin, London.

IVES, H.E. (1951). *Proc. Am. phil. Soc.* **95**, 125.
IVES, H.E. and STILLWELL, G.R. (1938). *J. opt. Soc. Am.* **28**, 215.

KESWANI, G.H. (1965). *Brit. J. Phil. Sci.,* **15**, 286, **16**, 19.
(1966). *Brit. J. Phil. Sci.,* **17**, 234.

LANCZOS, C. (1959). *Am. Scient.* **47**, 41.
LOEB, L.B. and ADAMS, A.S. (1933). *The development of physical thought.* J. Wiley and Sons, Chapman and Hall, London.

McCREA, W.H. (1951). *Nature, Lond.* **167**, 680.
(1967). *Nature, Lond.* **216**, 122.
MØLLER, C. (1952). *The theory of relativity.* Clarendon Press, Oxford.

POUND, R.V. and REBKA, G.A. JR (1960). *Phys. Rev. Lett.* **4**, 337.
POUND, R.V. and SNIDER, J.L. (1965). *Phys. Rev.* **140**, **3B**, 788.

RICHTMYER, F.K. and KENNARD, E.H. (1950). *Introduction to modern physics,* p. 121. McGraw Hill, London.
ROSSER, W.G.V. (1964). *An introduction to the theory of relativity.* Butterworth, London.
(1969). *Wykeham science series for schools.* Wykeham Publications (Taylor and Francis).
RUSSELL, B. (1927). *The analysis of matter.* Kegan Paul, Trench, Trubner, London.

SIVADJIAN, J. (1953). *Arch. Sci., Genève.* **6**, 191.
SYNGE, J.L. (1956). *Relativity, the special theory.* North Holland, Amsterdam.

TOLMAN, R.C. (1934). *Relativity, thermodynamics, and cosmology*. Clarendon Press, Oxford.

WHITROW, G.J. (1961). *The natural philosophy of time*. Nelson, London.

WHITTAKER, E. (1949). *From Euclid to Eddington*. Cambridge University Press.

(1953). *Eather and electricity*, Nelson, London.